THE GREAT SHIP
VASA

THE GREAT SHIP
VASA

WRITTEN AND DESIGNED BY

Greta Franzen

HASTINGS HOUSE ∽ NEW YORK

ISBN: 8038-2647-8

Library of Congress Catalog Card Number: 72-150019

PRINTED IN THE UNITED STATES OF AMERICA

PICTURE CREDITS

The right to reproduce the pictures in this book has been courteously granted by the following sources:

Anders Franzén and Norstedts/Bonniers, for pictures from *The Warship Vasa*, Anders Franzén, Norstedts/Bonniers, 1960, 1966, Stockholm: pages 24, 28 (left), 35, 41, 48, 49, 50, 52 (top), 54, 55.

Anders Franzén and the Stockholm *Expressen*: page 42.

Nicolai Kowarsky: page 17.

Gunnar Olofsson: page 79.

Swedish National Maritime Museum: pages 74, 75.

Swedish National Maritime Museum and Wasa Museum: pages 11, 23, 31, 40, 44, 46, 52 (bottom), 56, 57, 58, 60, 61, 62, 63, 64, 66 (top left, bottom left), 68, 70, 72, 76, 77 (left), 79, 81 (top left), 82, 86, 87, 89 (top).

The remaining pictures are copyrighted by the author: pages 13, 18, 27, 28 (right), 29, 30, 38-39, 66 (top right, bottom right), 67, 69, 77 (right), 81 (top right, middle, bottom), 83, 88, 89 (bottom), 90, 91, 92.

To Lt. William M. Lyons

ACKNOWLEDGMENTS

The author gratefully acknowledges the help of the following people whose cooperation made this book possible:

Anders Franzén, the discoverer of the *Vasa*, for opening the necessary doors, for giving me permission to use material from his own books and for granting the taped interview from which most of Chapter 3 was derived;

Bengt Ohrelius, public relations director of the Wasa Museum, for supplying me with most of the information on the court trial as well as other important facts and for allowing me *carte blanche* use of the museum's picture files;

Katarina Villner, his assistant, for her prompt and tireless replies to the countless questions that needed answers;

Lars Barkman for his most precise explanation of the preservation department which he heads.

[6]

CONTENTS

THE GREAT SHIP
VASA

1. *The Maiden Voyage*

THE sun rose on the tenth of August, 1628, spreading its rays of warm promise over the calm, blue harbor and the gleaming spires of the city of Stockholm. The residents were abroad early, even though it was the Sabbath, for this was the long-awaited day when the great ship, the *Vasa*, was to start on her maiden voyage. As the day moved lazily on toward afternoon, the southern hills and the numerous islands in the harbor and archipelago, as well as the quays, became dotted with people in a festive mood. Many had gathered with friends for a picnic or some other merriment while waiting for this important event.

The *Vasa*, named for the King of Sweden's family, the House of Vasa, was the newest and grandest of the fleet of warships of Gustavus II Adolphus.* In 1625 he had ordered a Dutchman, Henrik Hybertsson, and his brother, Arent Hybertsson de Groot, to build four ships, two small and two large, of which the *Vasa* was one of the larger. These ships, along with many others in his powerful navy, were needed by the King to carry men and supplies to the Polish province of Prussia, which linked Poland to the Baltic Sea. There he was embroiled in what came to be known as the Thirty Years' War (1618-1648). Sweden had already fought against Denmark, Poland and Russia and had acquired both Finland and Esthonia. Now, as the champion of the Lutherans who feared the advances of Catholicism into northern Germany, the King was locked in a deadly struggle with his cousin Sigismund, Catholic ruler of Poland.

His motives, however, were not purely religious — they were also political and economic. As a brilliant statesman and a military genius, he had raised Sweden from a backward country to one of the most powerful nations in Europe. This power he had no intention of losing. What is more, he intended to make the Baltic Sea into a Swedish lake. This would mean not only security from attack by foreign rulers but also increased trade. What better way to accomplish this end and also win the war than by having a great navy to transport men and supplies?

Hence the *Vasa*'s keel was laid in 1626 at the Royal Dockyard at Blasieholmen, then a small island in the harbor a few hundred yards from the Royal Castle. Launched in 1627, she had become the pride of the nation as progress reports reached the farthest provinces of the land. Surely, if the stories of her magnificence could be believed, few other ships could compare with this stately queen of the Baltic! It was told that she was decorated on the bow with a beautifully carved lion and on the stern with many and varied sculptured figures. Fierce lions' faces guarded each of her gun ports, glaring out to frighten off any potential enemy. With all these adornments, brightly painted and gilded, with tall masts

---

* He reigned from 1611 to 1632.

The Swedish monarchy and the countries across the Baltic Sea in the early 17th century, during the Thirty Years' War.

and enormous sails and with many flags and pennants flying, she would indeed present an awesome sight.

Of course, there were some who had mixed emotions about the glorious day. These were the wives, children and other relatives of the crew and soldiers. The latter were to come aboard farther along the archipelago, probably at Älvsnabben, where the ship was to await orders from the King. Little wonder that these people felt sad, for they knew that it might be years before they would see their loved ones again. Especially unhappy were those whose men

[13]

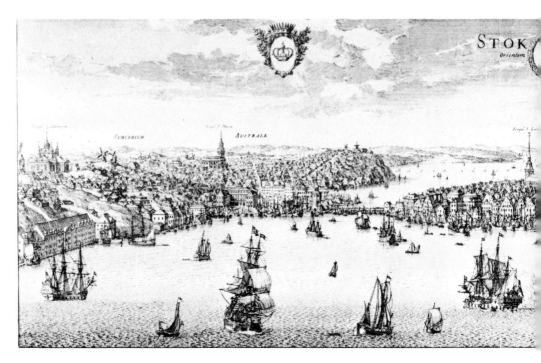

An old engraving by Villem Swidde, showing 17th century Stockholm. In the foreground, right, is Kastellholmen. Directly in back and to the left is

were neither professional sailors nor soldiers but just ordinary farmers, laborers or artisans. They had not been trained for a harsh life* at sea nor did they desire to fight a war that had already been long and costly both in supplies and loss of lives.

By 1628 the number of eligible men around Stockholm had been depleted since only five percent of the population lived in this urban area. For this reason the officials of the Crown had to ride out on horseback into remote provinces to conscript new recruits. Actually, it was a law in those times that one man from every twenty homesteads should volunteer for the King's service. However, if a man did not agree to serve of his own free will, the officers had orders to force him to join. In this manner, then, was the ship's contingent made up.

---

* Discipline was very strict and severe punishment was meted out for violations of the rules. A favorite penalty was tying a rope around the offender, throwing him overboard and pulling him from one side to the other underneath the boat.

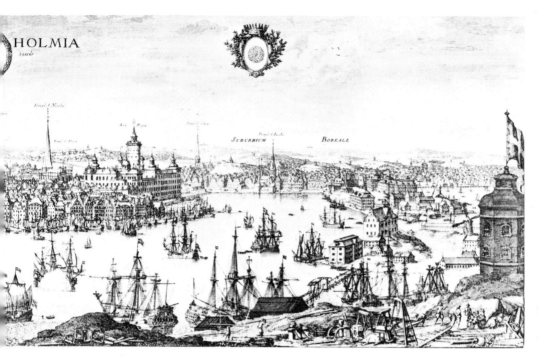

Blasieholmen, where the Vasa was built. In the center is "Old Town" with the castle and dock on its extreme right.

Now at last the great *Vasa* was waiting to be cast off from her dock below the Royal Castle at Tre Kronor Fort on the island, now called "Old Town." She had been towed there in the spring from her "birthplace" at Blasieholmen to take aboard her ballast, her armaments and provisions. This task had been completed ten days before on July 31. On board for the occasion were some of the sailors' wives and children, who, according to the custom of that time, were allowed to remain until ordered to disembark, probably at the spot where the soldiers were to join the ship.

Between three and four p.m. the bells from the Great Church of Stockholm pealed out the end of Vespers and, at a given signal, the *Vasa* was towed slowly from her berth. Majestically, the proud beauty of the seas moved past the quays of the city, heading south past the heights of Södermalm. Her flags and banners lifted gently in the slight breeze and her many gilded ornaments threw back the rays of the late afternoon sun to the thousands of cheering spectators on the hills and shores and in the boats in the harbor. Meanwhile, four sails had been set — the fore-topsail, the foresail, the main-topsail and the mizzen. A two-gun salute sounded the ship's farewell. Everything was going smoothly as she neared the island of Beckholmen. Suddenly, without warning, a violent gust of wind from the south-southwest filled her sails, causing her to list sharply to port. An abrupt silence from the onlookers was followed by cheers as she slowly righted herself. Then moments later she again listed, this time so far to port that water rushed in through the open gun ports. In a matter of minutes, down the great *Vasa* sank to the bottom in 110 feet of water!

The people all around stood frozen, their cheers turned to mute expressions of horror and shock at this unbelievable turn of events. A day that started with so much expectancy had ended in a national catastrophe. Surely, this was the shortest maiden voyage in all naval history!

The many smaller boats in the harbor, sailing close to the *Vasa* to bid her "Godspeed," managed to save some of the people in the confusion and hysteria that followed. The estimates of the number of dead varies, but it is believed some fifty individuals were drowned. Among them were several women and probably most of the seamen who manned the guns, since they were at their stations below deck. In those days the gunners were even expected to sleep beside their guns.

Among those to survive were Captain Söfring Hansson; Lieutenant Peter Gierdsson, the Rigging Officer; Erik Jönsson, the Chief Ordnance Officer; Jöran Matsson, the Sailing Master; and Per Bertilsson, the Chief Boatswain. On reaching shore, they were all arrested and imprisoned. Arrested also were the ship's builders.

The following day all these men were questioned at length by the Council of the Realm (the governing body), hastily assembled at the palace. On Tuesday a sorrowful letter, promising a full investigation, was sent to the King, a letter which was preserved and which later played a major role in pin-pointing the *Vasa*'s resting place.

On September 5th, a Court of Inquiry, appointed by the Council of the Realm, met at the palace to conduct the investiga-

Artist Nicolai Kowarsky's conception of the Vasa, *drawn in 1968. The four sails hoisted on her maiden voyage were: (A) mizzen (B) main-topsail (C) fore-topsail (D) foresail.*

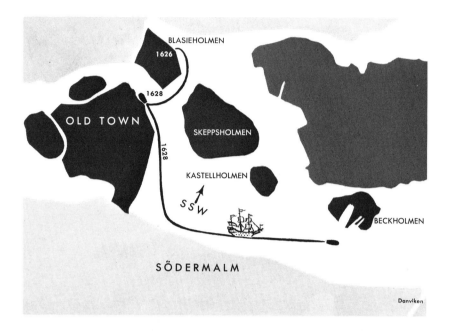

This map of Stockholm harbor shows the voyage of the Vasa *from her birthplace at Blasieholmen to the point where she foundered.*

tion. Composed of 17 members, all notable men — six of them Councillors of the Realm — it was headed by Admiral Carl Carlsson Gyllenhielm, a half-brother of the King. Now, one by one, the ship's officers and the builders were called before this august group and interrogated thoroughly. The members of the Court apparently were ignorant of the chain of command on the ship and the job for which each man was responsible, because they threw accusations at each witness indiscriminately.

Copies of parts of the records of this inquiry — the official record has not been found — have been preserved in the National Record Office.* From these one can piece together enough of the testimony to get a good picture of the proceedings. Through it all it was evident that the Court was out to find a scapegoat and tried to pin the blame on each man questioned. Punishment for the guilty would have been certain death, so each of the accused did

---

* Brought to light by the research of Georg Hafström, Swedish naval historian.

his utmost to prove his innocence by pleading ignorance or placing the blame on an officer higher up. The word "crank" — a nautical term meaning *unsteady, badly balanced* — appeared again and again in the testimony and the prosecutor seemed determined to prove that Captain Hansson had been aware that the ship was crank even before the start of the voyage.

One of the first to appear before the Court was Erik Jönsson, the Ordnance Master, who had been appointed by the King himself to command the fleet that the *Vasa* would have joined. When the prosecutor accused him of not having familiarized himself enough with a ship under his command to make sure that she had enough ballast, he replied that it was up to the Captain or Sailing Master to see to that. His own duty was the overseeing of the armaments and guns only, though he would take command during battles. He had, as a matter of fact, gone below and inspected the guns after the ship listed the first time, thinking that they might have become loose and rolled to port. Now he blurted out that the *Vasa* had listed earlier while she was still lying inshore — even before the sails had been set! However, everything had seemed secure and he thought nothing of that incident during the second list. When the ship started to keel over for the third time and seemed in danger of capsizing, he again had rushed down and ordered the guns dragged to starboard to counter the list. Water was even then pouring through the open gun ports and he had to move fast to save himself. On further questioning, he admitted that he had not trusted the ship — that in his opinion she was top-heavy and could possibly have capsized even without moving, as witness the first list. When asked if he thought more ballast should have been added, he countered with the question, "How could more ballast be added when the gun ports were dangerously low to the water already — only three and a half feet above the water line?"

As Peter Gierdsson had his turn before the Court he was accused of not informing the authorities that the ship was top-heavy and did not have adequate ballast. He answered that his only responsibility was the rigging. He knew nothing about shipbuilding nor what kind or how much ballast should be used. He never would have thought it possible for a ship to capsize in such a slight wind. His testimony did not add much toward solving the case. He did confirm, however, that the guns were secure.

Jöran Matsson, the Sailing Master, was a less reluctant witness. When charged by the Court of not stowing enough ballast, he insisted that he had filled the hold to capacity. He had personally checked to see that the ballast was sufficient and placed correctly. The answer to the next question put to him created a sensation. He was asked if he noticed, prior to departure, that the ship was top-heavy. He replied that even while the ship was still at the quay, he knew Captain Hansson had reported to Admiral Fleming that he thought she was crank! Furthermore, the two had conducted a stability test. They had ordered 30 men to run from one side of the ship to the other to see if she would tilt because of the moving weight. The result was that each time the men ran across, the *Vasa* tipped the depth of one plank. In other words, the first time she sank down the depth of one plank; the second time, two planks; the third time, three planks. It was then Admiral Fleming called a halt to the experiment. He had admitted that the ship would have keeled over if they had made more runs! What is more, Fleming could have told the King, himself, since Gustavus II Adolphus had been home at this time.

Matsson, continuing his testimony, stated that even before the test he had informed the Admiral that the ship was too narrow-bottomed and did not have enough belly to hold the right amount of ballast. The Admiral had retorted that she was carrying too much ballast as it was, since the gun ports were now too close to the water. To this he had added that Matsson was not to worry since the shipwright had built ships before. Finally, when asked if the *Vasa* had too much sail for the prevailing weather conditions, Matsson answered by asking the question, "If the ship capsized in sheltered waters with only four sails, how would she have fared out at sea with all sails set?"

The only surviving officer left to testify, except Captain Hansson, was Per Bertilsson, the Chief Boatswain. Now that the other naval witnesses had cleared themselves, the Court attempted to put the blame on him by accusing him of being drunk and neglecting his duty. He saved himself, however, by swearing that he was not drunk and had that very day taken Holy Communion before the ship had sailed.

Now the shipbuilders were asked to appear before the Court. When Hein Jacobsson of Holland was called to the stand, he was asked why he had built such a badly-designed ship. He was ready with an easy answer. He had only finished the job that the Master Shipwright Henrik Hybertsson had started. Hybertsson, commissioned by King Gustavus II Adolphus, had died in 1627, after laying the keel and starting the framework. The original plans, called a "sert," had been approved by the King. There were no naval architectural plans made in those days and a sert consisted merely of a list of the important dimensions. This also served as a contract to build. Jacobsson claimed he had only followed the original specifications as approved by the King and thus neatly placed the blame in the King's lap. Under questioning, he admitted, however, that he had not built the ship exactly to specifications. Instead he had increased the width of the hull by 17 inches because he felt that the ship was too narrow and sharp and did not have enough "belly"

A bronze bust of King Gustavus II Adolphus sculpted by Hans von de Putt in 1632.

to suit him! Hybertsson's brother, Arent Hybertsson de Groot, a merchant, was also questioned since he had been involved with him in the sert. He too fell back on the defense that the *Vasa* was well built according to the specifications submitted to his sovereign. In his estimation she would not have capsized if there had been sufficient ballast.

Captain Hansson, now called to testify, objected to de Groot's statement, reiterating that the gun ports were already too close to the water. Furthermore, said Captain Hansson, Hein Jacobsson had on one occasion boasted to him that the ship was "as tight as St. Peter and would be able to sail even without ballast." Jacobsson denied saying this and insisted that if the *Vasa* had had more ballast she would not have capsized.

After many more such recriminations and sparring between the builders and Captain Hansson,* the Court adjourned without coming to any decision as to where the fault lay, and the men were freed. Admiral Fleming was never asked to testify, although the facts as presented must have made each member of the prosecution realize who undoubtedly should have shouldered the blame. But what man would risk censuring the Admiral publicly, when such accusation would point directly to the King? After all, the plans had met with royal approval in the first place.

And so the people of Sweden could only wonder about the reasons for this catastrophe and argue among themselves about these puzzling questions:

1. Was the *Vasa* built too narrow and sharp, without enough belly as Matsson insisted?
2. Was she top-heavy?
3. Were the lower gun ports too close to the water level?
4. Were the guns at fault — too heavy and misplaced?
5. Would there ever be an answer to this mystery?

---

* Surprisingly, there is very little testimony attributed to Captain Hansson in the known record. One can only assume that he must have been questioned at length and his evidence lost with the missing parts of the record.

WOOD SCULPTURE • WINGED HEAD

2. The Early Salvagers

THE large hulk of the *Vasa* had hardly settled in her unexpected berth before the authorities began wondering about salvage possibilities. After all, this treasure, the product of almost three years' work could not possibly be allowed to remain at the bottom of the harbor. Besides, the King had to be assured that not only would those responsible for this disaster be brought to trial and punished, but also that every effort would be made to return the ship to His Majesty's service as soon as possible.

In every age there have been adventurers — men who have risked their lives because of the lure of wealth or the challenge of a death-defying act. The salvage of the *Vasa* promised both riches and danger. So they came from near and far, men willing to take a chance at being successful in a salvage operation of this awesome magnitude. How they could even hope to accomplish such a task with primitive tools and without diving suits* of any kind to protect them, is beyond the imagination today in the light of our modern underwater technology and its sophisticated equipment.

* The first crude diving-suits appeared in the 1880's.

Furthermore, these men were not divers as we know them. They were merely good swimmers who could stay below longer than the average person. Even so, their diving must have been done in short, frantic plunges as they investigated the position of the ship and the possibilities of attaching ropes to it. All their probing had to be done in almost complete darkness near the harbor bottom since the sun's

An old print showing Swedish implements used in salvage work. From Konsten att lefwa under vatn (The Art of Living under Water), *by Mårten Triewald, Stockholm, 1734.*

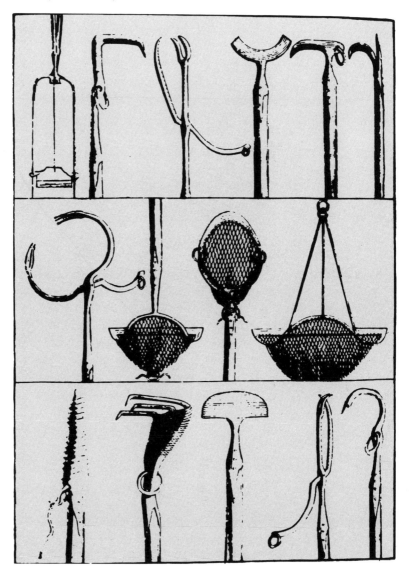

rays could not reach very far into these murky waters. The temperature of the water must have been very uncomfortable, even paralyzing, for it is always extremely cold at that depth.

The first to attempt this feat of salvage was an English engineer, Ian Bulmer,* who happened to be in Stockholm at the time of the disaster. According to a letter sent to the King, he had been authorized by the Council of the Realm, three days after the *Vasa* sank, to be the official salvage operator. He was to be paid only upon completion of the job. However, he was only able to get the ship back on an even keel, so that her main mast was sticking straight up out of the water instead of at an angle which had resulted from her sinking while listing to port. This was a remarkable accomplishment in itself and how he managed even this is a mystery, for no record was kept, or, at least, none has been found. One theory is that ropes were twined around the masts and attached to horses on shore that pulled the ship to the upright position.

Ian Bulmer continued for some time in his attempts to raise the *Vasa* but was gradually edged out by others — Dutch, French and Germans, all of whom failed. In 1629 the Swedish Navy made a stab at rescue work, but it also gave up because of the inadequacy of the equipment available. Slowly the *Vasa* and her surroundings became a regular graveyard for anchors, winches, grapplers and other hardware used by her would-be salvagers, and she lay almost forgotten for a number of years.

The next person of any stature to attempt the salvage was an ex-officer of the Swedish Army, Lieutenant-Colonel Hans Albrecht von Treileben, who came on the *Vasa* scene in 1663. Born in 1625, only three years before the fatal event, he was too young to remember it himself, but developed an interest in the ship in later life. After contracting a serious disease while fighting in Poland, von Treileben retired from the army in 1655, remaining on the continent where he studied the art of diving. It was he who brought back the diving bell† to Sweden. Although used elsewhere in Europe

---

* He called himself "Engineer to His Majesty, the King of England."

† The first diving bell has traditionally been attributed to Roger Bacon, c. 1250, but no records have been found to substantiate this.

for many years, it introduced to the Swedes a new concept in underwater exploring. In 1658 von Treileben had used it for the first time to bring to the surface several cannon belonging to the *Sancta Sophia*, a Danish ship that had sunk off Gothenburg, on the west coast of Sweden, also in 110 feet of water. For some years after, he kept up his salvaging activities in both Swedish and foreign waters.

Then in 1663 he joined forces with a German salvage expert by the name of Andreas Peckell, who had come to Stockholm because he, too, was interested in the *Vasa*. He had heard about the ship while fighting with the Swedish Army in Poland. Peckell's idea was to dismantle the ship from the top and bring up only the valuable bronze cannon rather than try to salvage the ship as a whole. Before this skilled combine could start its operation, however, von Treileben had to carry on some shrewd negotiations to secure the royal salvaging privileges, held since 1652 by Colonel Alexander Forbes, a Scotsman serving in the Swedish Army, who never himself did any salvaging.

With this obstacle cleared away, work was finally begun in the fall of 1663. A Scotch diving expert, Jacob Maule, who had worked on the *Sancta Sophia*, joined the team. Weeks were spent in clearing off the debris left by all the predecessors so that the ship could be approached. Finally, on April 1, 1664, the first gun was brought to the surface. The diving continued for the rest of that year, when the weather allowed, and far into 1665. Fifty-two more cannon were salvaged. These, according to the records, were sold to Germany.

This incredible operation undoubtedly owes its success to von Treileben's diving bell. A description of this bell and the diver's clothing can be found in a book written by an Italian priest and explorer, Francesco Negri.* While on his way to the North Cape in October, 1663, he stopped off in Stockholm to investigate a story that had interested him. He had heard that someone had discovered how to descend comfortably and without risk into the depths of the sea. To satisfy his own curiosity, Negri, with a friend, went

---

* *Viaggio Settentrionale* (Padova, 1700)

Front view of an open gun port, showing lion's face and gun. This is one of two guns on display in the Vasa Museum yard. Each gun weighs about 3000 lbs. The ship's hull is a modern reconstruction.

Side view of gun and gun port.

aboard a large boat from which von Treileben's diving operations were carried on.

According to his account, the diver first of all put on boots that came above the knees and were fastened there by rope tied around the legs. Rough leather stockings were pulled over these boots and a leather jerkin over the diver's shoulders and chest. Weights were then attached to a metal ring around his waist and a metal ring above each knee. On his head, the diver wore a cloth hood. With the exception of the headgear, this outfit was water-tight and fairly pliable to allow for easy movement.

After he was dressed, the diver descended to a log raft on which the diving bell stood. This bell, resembling a church bell,

LEFT: *Francesco Negri's drawing of the diving bell from his book,* Viaggio Settentrionale.

RIGHT: *A modern copy of von Treileben's diving bell, standing in the Vasa Museum yard. Designed by Anders Franzén from Francesco Negri's description, it was demonstrated in 1960. Lt. Commander Bo Cassel descended in the bell to the Vasa and broadcast his impressions through microphones linked to both radio and television.*

A contemporary artist's conception of a diver retrieving a gun in 1663. This poor chap is not wearing the boots described by Negri. At the right is a barrel of extra air connected to the bell by a hose. This innovation was made after salvage work had begun.

was made of lead and was about four feet, two inches high, with a rope pulled through a ring on top. When the bell was raised by means of a block and tackle, the diver was able to enter it. He stood on a lead platform that hung twenty inches below the bell and was attached to it by a rope pulled through a hole in each corner of the platform and through corresponding holes in the lower rim of the bell. The diver's only piece of hand equipment was a six-foot wooden stave which had an iron boat hook on one end.

As the bell was lowered into the water, a certain amount of air* was compressed by the cold water and trapped at the top. This air pocket allowed the diver to be submerged for about fifteen

* One quarter of the total volume of the bell for the intended depth.

minutes, even though the water was now up to his neck. His only means of communication was a rope which went out from under the rim of the bell to the surface and which the diver could pull to signal the watchers on the raft.

It is hard to believe that these men with their primitive equipment could accomplish the astounding feat of recovering the guns, but they did. For each gun removed from the lower gun decks had to be pulled out through the gun ports by means of the boat hook. Then a rope had to be wound around the gun and secured before it could be hoisted to the surface. One wonders how the diver could see to do any of this, but it was recorded by Negri that the rays of the sun penetrated the water for at least a short distance and the diver could see fairly well looking straight downward. This, of course, was in the 1660's before the silt and dirt of the centuries had settled over the *Vasa*. Added to these problems, he still had the cold to combat. Negri reported that when the diver reappeared on the surface, he was shaking as if he had the ague.

With the 53 cannon recovered, the team of von Treileben and Peckell* was dissolved. After that, except for the salvage of another gun in 1683,† the *Vasa* was abandoned again to her fate and forgotten, this time for more than two centuries.

* Von Treileben, Peckell, Forbes and Maule went through several lengthy court cases suing each other. Most of the facts in the chapter are based on the court records.
† A man named Liverton was the salvager. There is little information available about his operation or the fate of the gun.

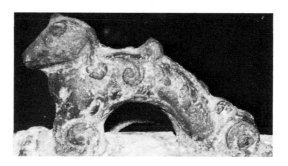

Detail from Vasa *gun*

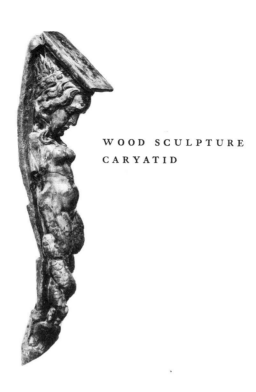

WOOD SCULPTURE
CARYATID

3. *The Discovery*

IN the summer of 1920, an incident became a link in the eventual rediscovery of the *Vasa*. Erik Nordström, a fisherman, had dropped his anchor in a spot off Landsort at the opening of the Stockholm Archipelago. When ready to move on, he found that the anchor would not budge. By chance, someone aboard a salvage vessel a short distance away noticed his predicament and cruised over to investigate. One of its divers volunteered to go down and dislodge the anchor if he were given a bottle of cognac as his reward. Although this was a high price for a poor fisherman to pay, he agreed to it, for he had lost several anchors in the same area and was curious about the reason. Nordström got his anchor back all right, but a far greater treasure was found by the diver. The anchor had gotten stuck in the wreck of an old warship.

Seven bronze guns were eventually retrieved by the divers and sold to the government for $12,000. These are now in the National Maritime Museum.

From the inscription on these guns, Professor Nils Ahnlund, the eminent Swedish historian — an expert on the 16th and 17th centuries — was able to identify the ship as the *Riksnyckeln* (*The Key of the Realm*). This had been another warship in King Gustavus II Adolphus' fleet. It was recalled then that 1628 had been an especially disastrous year for the Swedes. The records showed that this ship had been sunk in September of that year after smashing into some rocks. She was on her way home with a load of soldiers wounded in the war raging in Germany.

While doing research on this ship in the National Archives, Professor Ahnlund stumbled on information about the location of the *Vasa* which was later to play a part in her recovery. Some flurried stabs were made at this time to find her. Even a magician with a gold divining rod circled the area! All attempts proved futile, however, and interest in the *Vasa* waned when it was assumed she had disintegrated.

Meanwhile, Anders Franzén, who was later to discover the resting place of the *Vasa*, was growing up in Stockholm. Born there in 1918, he spent his childhood summers on Dalarö, an island in the Stockholm Archipelago. With his father, Dr. Anders Franzén, he spent many long summer days cruising around the nearby islands and the harbor itself. One of his father's hobbies was history and Dr. Franzén would tell his son about the important events that had transpired in that area, such as Danish battles or encounters with the Russians, and he would point out the spots where ships had sunk many years before.

Here at Dalarö, the father related, a ship of the Swedish Navy, the *Riksäpplet* (*The Apple of the Realm*)* had broken loose from her moorings, crashed into a rocky island and sunk in 1676. The position of this vessel had never been forgotten because the story of its foundering had been passed down from father to son in this fishing community of Dalarö. From time to time different people had tried to salvage the remains. For instance, in the 1920's, some black oak had been brought to the surface from which furniture had been made for the King. On one of these salvage operations a diver had died. Since the Franzéns lived in the area and Anders' father was the only doctor nearby, he was called to the scene to verify the death.

In this way the family got to know these people and received as gifts some of the objects from the *Riksäpplet*. Among them were the wheels of a gun-carriage — still in Anders' possession. He remembered in later life that he was astonished at the good condition of the wood — a fact that he then thought no more about, but was later to recall.

Although Anders, during these outings, was more interested in running their boat than in listening to the tales of old naval battles, he absorbed, in spite of himself, a good deal of the local history. The stories of sunken ships especially intrigued him and, with a young boy's curiosity, he began to wonder what was left of these old wrecks. Why, there must be thousands of treasures lying at the bottom, just waiting for someone to come along and discover them!

Time went by and then another incident occurred which later proved important to Anders. In his late teens, shortly before World War II, he took a boat trip with his father through the Göta Canal to the west coast of Sweden. There he saw an old wreck that had been salvaged. He observed that the wood was very spongy. This, his father told him, was due to the ship worm, the *teredo navalis*, which had attacked the wood. This puzzled him, for he remembered the good condition of the wood in the *Riksäpplet* gun wheels. Why should salvaged wood on the west coast be in such poor con-

* Or *The Orb of the Realm* (Orbus Terrarum)

dition while wood retrieved from the Baltic was practically like new? To satisfy his curiosity, he began searching for a reason and discovered there were no *teredos* in the Baltic. The *teredo* could not exist in waters that had a low salt content, such as the Baltic Sea.

Slowly, recalling his childhood dreams, he became convinced they might become a reality. These waters must surely be hiding a unique and extraordinary treasure-trove, for the Baltic was probably the only sea where large sailing ships had been operating and where wood could survive the onslaught of natural enemies. The fact that the Baltic had no *teredos* had been known by marine biologists and other scholars, but Anders Franzén was the first to realize its possible connection with foundered ships. Now, spurred on by this exciting theory, Anders renewed his determination to search for sunken treasure.

Then came World War II, and Anders entered the Royal University of Technology in Stockholm to become a naval architect. He never worked as one, however, for he invented a device used in the oil business and has been connected with this field ever since, being at the present time a Swedish Navy petroleum engineer. While at the university, he studied Swedish naval history and delved through the naval archives, where he discovered the names of many old wrecks. He made a list of twelve* dating back to the 16th and 17th centuries that he intended some day to investigate.

He singled out those of the 16th and 17th centuries for very good reasons. In his estimation, it would be of no historical importance to salvage a wreck unless it could be identified. It would be very difficult to identify any ships built before the 16th century,

---

* The *Riksvasa*, (*The Vasa Kingdom*), 1623, the *Vasa*, 1628, the *Riksnyckeln* (*The Key of the Realm*), 1628, the *Riksäpplet* (*The Orb of the Realm*), 1676, the *Gröne Jägaren* (*The Green Huntsman*), 1676, have been located and investigated. The *Lybska Svan* (*The Swan of Lybeck*), 1525, the *Lybska Örn* (*The Eagle of Lybeck*), 1576, the *Mars*, 1564, the *Svärdet* (*The Sword*), 1676, the *Kronan* (*The Crown*), 1676, and the *Västervik* (*The West Bay*), 1676, have not been found but their approximate locations are known from the old records.

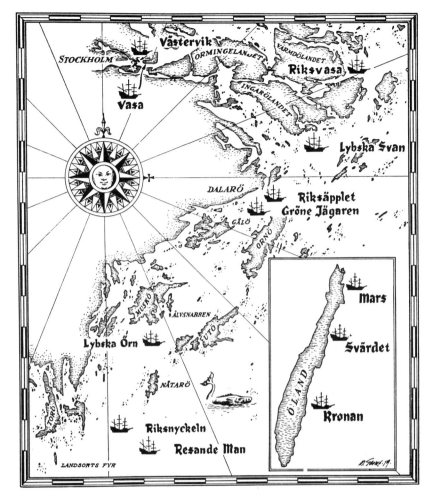

This map of the Stockholm Archipelago, drawn in 1959 by Magnus Gerne, shows the locations of the 12 wrecks on Anders Franzén's list. Note Älvsnabben, where supposedly the soldiers were to board the Vasa.

should such be found, for there would be little information available about them. The archives were not organized until 1523* and the first ship recorded by name dated back to the early part of that century. On the other hand, ships dating from the 18th century

---

* This was under the instigation of Gustav Vasa, who united Sweden and in that year became her first real king.

did not interest Anders, since by this time not only were their names recorded but also their plans. Bringing one of them to the surface would add little to marine archeological history and would not be worth the cost.

Later, Anders took a course in diving given to naval officers so that they could in turn instruct divers under their command. They practiced their diving at Dalarö and investigated the remains of the *Riksäpplet*, the ship that Anders had known about since childhood. It was also on his list. Then, and later as an amateur marine archeologist in cooperation with the National Maritime Museum, he helped to locate and examine others. The more he worked with them the more he realized that of all the twelve ships, the *Vasa*, so elusive, would be the most intriguing to locate. She would be the most rewarding to bring to the surface and well worth the cost, too. This was true since she once was a large and magnificent ship; she was new when she sank and could be in good condition — that is, if his theory of the absence of the *teredo* was correct. Also, she had sunk in a protected harbor, close to the naval dockyards where salvage operations would be feasible.

During World War II he was in a position to receive a great deal of material about frogmen (called "fighting divers" by German newspapers) from both sides of the war. The divers were shown in light outfits, carrying their oxygen or air supplies on their backs. Still thinking of underwater research, he realized how useful such an outfit would be in searching for old wrecks and investigating them. Two or three people could go out in a motorboat and search freely without utilizing the heavy barges and costly equipment of conventional heavy, helmeted divers. He decided to look into this possibility more fully after the war.

As the years passed by, Anders' determination to search for the *Vasa* grew stronger. He studied the archives and read all the documents pertaining to that era that he could find. Year after year in his spare time, he searched for any facts that might be useful. To decipher the old handwriting, he even learned "old Swedish."

One day he found a source relating to the place where the

Vasa had sunk. It was vague — ". . . behind Lustholmen, this side of Blockhusudden, by Danviken," and covered a wide expanse of water.* Excited by this information, he thought he had the answer to the general location, and in the summer of 1953 he began his first actual search. He spent hours circling the area in his motorboat, towing heavy grappling irons. Again and again they would get caught, he would haul in the lines, and would be rewarded with nothing but the usual jetsam found on harbor bottoms all over the world: old motors, drowned animals, bicycles, and so on. If the line grew taut and he could not dislodge it right away, he would drop his core sampler into the depths. This was a device he had invented which, if it hit wood, would bite into it and bring a sample to the surface. He hoped that it would some day bring up black oak. Oak would indicate a big ship; black oak, a big, old ship. But the summer passed and with it the hope of finding even a meager trace of his quarry. The old record must have been wrong.

During the following autumn and winter he renewed his study in the archives. More reading and more sifting of pertinent facts followed. Then Anders came upon an old map drawn by a historian, which showed the *Vasa*'s position according to his theory. A cross had been marked near Stadsgårdskajen. Grasping at any straw, during the next summer (1954) he checked out this location. Again and again, every chance he had, he cruised around this part of the harbor but without any luck. The same thing happened in 1955.

Undaunted, Anders kept up his search for another clue to the location of the ship. Besides delving into old documents, he consulted many historians who might shed light on his quest. One of these scholars was Professor Ahnlund who had identified the guns of the *Riksnyckeln*. It was he who told Anders what he had stumbled on so many years before — the probable location of the *Vasa*. The

* See map on page 18. Lustholmen is now called Skeppsholmen, and Blockhusudden was probably Beckholmen.

This contemporary view of Stockholm harbor, taken from Södermalm, shows the great expanse of water Anders Franzén covered in his search. In the lower left hand corner are the quays of "Old Town" which the Vasa

clue, however, was in the letter sent to King Gustavus II Adolphus by the Council of the Realm in 1628 which was found by Anders in the archives.

Having mastered its old language, he finally learned its secret. It read in part " . . . it so happened that she came no further than to Beckholmsudden ere she went to the bottom. . . ." This was the information he needed. As early in 1956 as possible, he began systematically to comb this new area in his boat. The weeks passed by without any hits and it seemed as if yet another summer would be spent aimlessly circling the harbor. Then in August, 1956, it happened! His core sampler finally brought up a piece of black oak. What excitement he felt! Could this be the end of the chase? Was his theory finally to be realized?

Now the time spent in learning how to dive as a naval officer paid a dividend. Although he himself did not dive any more, Anders

passed. At left center is the island, Skeppsholmen, connected to Blasieholmen, at left, and Kastellholmen, at right, by bridges. The Vasa Museum is in the center. Above the crane at right is the spot where the Vasa foundered.

had kept his friendships with many of the personnel at the Navy Diving School. It was not a difficult task to persuade them to do their practice diving at the spot where he had brought up the oak sample.

Thus, that same month, with Anders Franzén and other divers manning the boat, the chief diver, Per Edvin Fälting, chose to be the first one to go down. When he reached bottom, Fälting reported to Anders over the loudspeaker that he was standing in mud up to his chest; he could see nothing; he could feel nothing. Anders told him to come up and they would try another spot. Just when Fälting was making himself lighter to ascend, he reported that he felt with his fingertips what he thought might be a wooden wall. After working his way closer, he said that it was a big wall and it could very well be the side of a big ship. While the men above waited in hushed expectancy, the diver slowly climbed upwards.

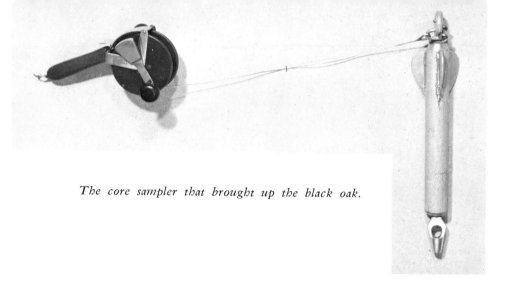

The core sampler that brought up the black oak.

Suddenly came the voice of the diver saying that he was standing in a square hole. Anders was sure now that this was an old warship, for the square hole must be a cannon port. Again came the voice, reporting that the diver had felt another square hole up to the left of the first.

Now Anders was positive that it was the *Vasa*. This had to be a three-decker because of its two covered gun decks, indicated by the zig-zag placement of the cannon ports. The *Vasa* was the only three-decker that had sunk in this harbor. At last his treasure was within reach! And Per Edvin Fälting became the first human to board the *Vasa* in almost three centuries.

However, the most difficult work lay ahead. Now that he had discovered the *Vasa*, Anders found that the prospect of bringing her to the surface was not very promising. The naval divers continued working around the ship, taking measurements and bringing up loose objects. The news of the find was reported in the press and brought world-wide interest. Local curiosity was aroused too, but little financial help was offered. Both the National Maritime Museum and the National Historic Museum, under whose jurisdiction such a find would ordinarily fall, were reluctant to do anything at this time, because their budgets had been planned for other endeavors.

Undaunted, now that he was so close to his prize, Anders decided to appeal directly to King Gustav VI Adolph, himself. The King was enthusiastic and wrote Anders a letter expressing his interest and stating that he hoped the salvage work could get started. He asked Anders to give him a report as soon as the plans had been formulated. This was the ammunition Anders needed, for it gave his project status. Armed with the letter, he was able to open many doors previously closed to him.

Now Anders made use of a clever bit of strategy. He requested from the Commander-in-Chief of the Swedish Navy that naval divers be officially assigned to the project, provided that the Neptun Salvage Company* could be persuaded to supply the lifting equipment. Then he went to the Neptun Salvage Company and asked for its cooperation if he could get the Navy to do the actual diving. In this way he interlocked the whole operation so that neither party could back out of its promise.

This would be a costly experiment for the Neptun Salvage Company. In terms of money, if this salvage operation had been

* A subsidiary of the Broström Shipping Company.

Diver Per Edvin Fälting coming out of the water after identifying the **Vasa**.

Anders Franzén with a wooden mermaid from the Vasa, *August, 1959.*

commissioned the probable fee would have been a half million dollars. The company was able to make this contribution, however, because often its equipment lay idle for months, maybe years, between salvage jobs. This happened to be a quiet period. Understandably, if a lucrative assignment should come along, the Neptun Salvage Company had the privilege of stopping all work on the *Vasa* until free again.

In Anders Franzén's estimation, though, the real cost to the Neptun Salvage Company was not its contribution of equipment but the risk involved. Should the ship be brought up in one piece, everybody would say that the company was fantastic. But should the ship come apart, the company's reputation would be ruined or at least badly hurt.

Everyone, of course, did not share Anders Franzén's enthusiasm for the *Vasa* nor have the vision to realize what a remarkable find she was. Many people doubted that she could be raised. Even if she could, what good would it be — just another way of spending money foolishly! Fortunately, enough individuals finally became interested in the project, and in the spring of 1957, the provisional *Vasa* Committee was formed with Commodore Edward Clason as chairman. Its task was to weigh the possibilities of raising the ship and solve the financial problems such a large undertaking would entail.

In the meantime, Anders traveled to the United States to study salvage operations there. He went to Norway to find out at first hand how the Viking ships had been reclaimed and he also went to Rome to learn as much as he could about the preservation methods for antiquities of all kinds. Being a technical man, he would give up the project of raising the *Vasa*, no matter how much that would sadden him personally, should it prove entirely unrealistic. Satisfied that there was a good chance of success, he decided to proceed with "Operation Vasa."

For Anders, one of the most exciting parts was over — arriving at his theory about the *teredos*; getting the idea that the *Vasa* might therefore be intact; developing this idea over the years with study and research; pursuing his quest alone; bringing up that first telltale piece of oak; and finally experiencing the triumphant moment when he had found the great *Vasa*.

But now the most dangerous part was about to begin.

WOOD SCULPTURE • TRITONS WITH SHIELD

4. The Salvage

WHEN word got around that the *Vasa* had been found and that a committee had been formed to study salvage possibilities, naturally many suggestions were made by well-meaning individuals from all over the world. In February, 1958, the Vasa Committee, after much consideration, finished its study, concluding that there was a good chance of salvaging the ship intact and recommending that she be brought up by conventional salvage methods. The plan was that she should be lifted by easy stages from her present position in 110 feet of water to a site of about 50 feet in depth and there remain submerged until further plans for the last stages of salvage could be formulated.

In the meantime, during the summer of 1957, the Swedish Navy divers had begun exploratory dives with the cooperation of the Neptun Salvage Company, as had been planned. Their equipment consisted of two lifting pontoons, *Oden* and *Frigg*,* the salvage ships *Ajax* and *Atlas*, and a lifting craft, the *Sleipner*, assisted by the Swedish Naval diving ship, the *Belos*.

The extent of damage to the ship was now disclosed. Imbedded in preservative slime and clay up to her water line, her whole top deck had been torn off. Her stern castle was almost entirely destroyed and the tops of the masts were completely gone. The lower foremast had been found standing in 1956 by diver Fälting, but had been removed to the surface for the divers' safety.

The destruction of the top deck was caused by the 17th century salvagers and the numerous anchors of ships passing over the forgotten wreck through the centuries. Many of these were subsequently found in her. Brought to the surface they were later placed around the Vasa Museum grounds.

In the course of these inspection trips, the divers found many wooden sculptures, which fortunately had fallen off as the bolts that held them had rusted. Hence, through the years, they had been naturally preserved in the mud and clay. The first brought to the surface was a lion's head from one of the gun ports that became a symbol or mascot for the whole project. Besides the sculptures, which will be described in Chapter Six, a number of loose construction pieces were surfaced prior to the initial lifting operation, such as the lower main mast, the thirty-foot rudder, the leonine figure-head and many more objects that either might have hindered the underwater operation or become damaged further in its progress.

A 24-pound bronze gun was brought up in September, 1958, by means of a pontoon crane. It took a whole day for the divers to disengage the gun from its position, bring it out through the gun port, and lift it to the surface with the aid of all known modern equipment. In the light of this, one cannot help wondering how von Treileben and his crew with their primitive tools could have extracted from the depths those 70 tons of cannon, the removal of

---

* Named after the one-eyed Norse god and his wife.

The first of the three remaining guns of the Vasa, *brought to the surface in September, 1958, by a pontoon crane.*

which, because of their weight, probably helped save the *Vasa* from greater damage.

When the immediate area around the ship had been cleared, the divers began the preliminary operations. The plan was to dig six tunnels under her keel through which cables would be drawn and attached to the surface pontoons, thereby forming a cradle that would lift the ship. The use of frogmen was early proved impractical. Although they had greater freedom of movement, they were hampered by the lack of visibility at the muddy bottom, and their light equipment was not sturdy enough for the heavy work involved.

Instead each diver was dressed in the latest, conventional heavy diving gear. He was equipped with an air hose and a safety line that was also combined with a telephone cable, connected to a loud-speaker in the diving boat. His most important piece of working equipment was the Zetterström* water jet, with which he did the actual digging. This was a nozzle on the business end of a hose through which a powerful, recoiless jet stream of water was forced, cutting through the mud and clay. This mucky matter, in turn, was sucked up by an airlift and brought to the boats on the surface to be filtered for small objects. The airlift was simply a six-inch rubber hose, connected to a smaller hose on its outside through which compressed air was forced downward from the surface. When it entered the larger tube, it roared upward at high speed, bringing with it water and the loose dirt.

During this tunneling operation, the diver was in constant danger. Working blindly in complete darkness as he crawled under the hull flat on his belly, surrounded by the cold, slimy mud, he knew that the tunnel might collapse at any moment, bringing instant death. What is more, he could get his airline fouled and suffocate before he could summon help, or his suit could get torn and he would drown before he could back out from under the hull. He realized, too, that right over his head was the ship's ballast of many

---

* This device was invented by Arne Zetterström, who broke the world's deep-diving record in 1945.

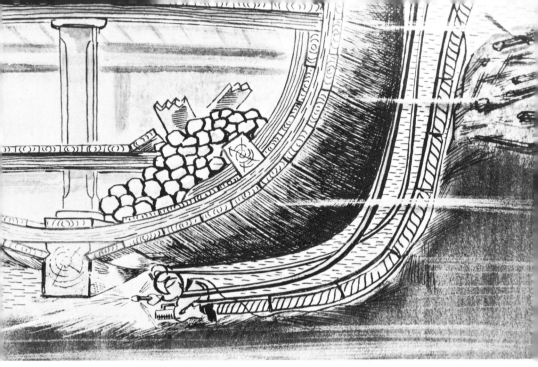

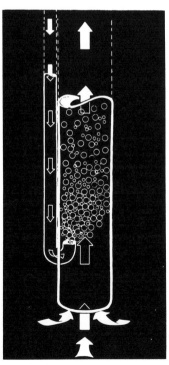

Artist Bengt Wallén's sketch of the tunneling. The diver is holding the Zetterström water jet. Between his knees is the airlift and connected to his waist are the combined safety line and telephone cable. The air hose is encircling his shoulder.

The airlift.

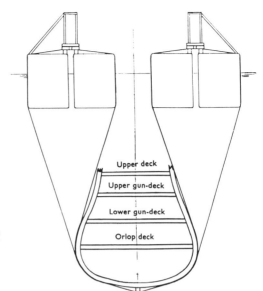

A cross-section of the lifting operation, showing how the wire cradle was formed between the pontoons.

Upper deck

Upper gun-deck

Lower gun-deck

Orlop deck

tons of stone, still lumped on the port side where it had slid on that fatal day. How strong could a wooden ship's bottom be after a 300-odd years' sojourn in the deep? Only the most fearless and courageous of men would tackle this job.

Such was the crew* headed by Chief Diver Per Edvin Fälting, without whom the *Vasa* could never have been raised. Month after month, during favorable weather, in the summers of 1957, 1958 and 1959, the digging went on. It was very slow going because of constant interruption by the traffic shuttling in and out of Gustav V dock. The salvage ships blocked the entrance for all but the smallest boats and had to be moved frequently. Finally, on August 20, 1959, the six tunnels, each nearly 80 feet long, were completed; two wire cables had been pulled through each one, forming a cradle. They were drawn taut and locked across the decks of the pontoons floating on the surface. The *Vasa* was ready for her first lift!

* They were Sven-Olof Nyberg, Stig Friberg, Lennart Carlbom and Ragnar Jansson, all petty officers in the Swedish Navy.

Artist's sketch of the Vasa *after the first lift.*

Now came the test. *Oden* and *Frigg* had been filled with water so that they were almost entirely under the surface. Slowly they were emptied so as not to create any suction under the hull. This would have made it impossible to lift her. As the water was being pumped out, the pontoons raised themselves out of the water, hopefully pulling up the ship by the wire cradle at the same time. But how could they be sure this plan would work? While the weary crew waited topside, hardly daring to move, a diver was sent down to investigate. When his voice shouted over the loudspeakers that "the *Vasa* has lifted 18 inches" cheers and applause broke out in the tremendous relief of the moment. Back to work went the crew. The ship was first turned 180° so that her bow faced the island of Kastellholmen, her destination. Then she was slowly towed about 40 yards before work was halted for the day. That night she rested on the bottom in her steel cradle.

In the next four days, two more lifts were accomplished and the ship was moved forward 230 yards. Now the salvage crew realized it would be easier to tow the *Vasa* by the stern. Being pulled bow first, as she had been, the wider front end had not only kicked up more dirt but also had settled farther down in the mud at the end of a lift. Finally all seemed well for an easy recovery. Then on August 25th trouble developed. The forward cable had slipped off the hull and became entangled. This could have been a costly accident but fortunately the cable was straightened out and pulled back under the hull. In spite of this delay, the ship was moved forward another 55 yards and upwards to a depth of 15 fathoms. On August 27th, heavy winds made the salvaging difficult and she slid back to the spot she was on August 25th. What is more, a ten-foot wall of clay now blocked her progress. On the next day, however, the best lift of all occurred. The ship was hoisted over this bank and moved forward 50 yards and upwards to a depth of 13 fathoms. Gradually, with adjustments now and then, on September 16th the *Vasa* was brought to a place outside Kastellholmen after 18 lifts. Here she was to lie at a depth of nine fathoms — for two winters, as it turned out. When the work stopped on this day for the rest of the year, everybody involved was both happy and

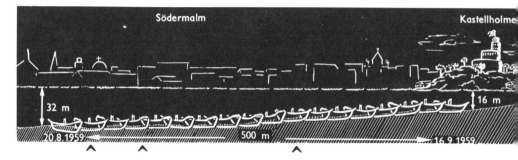

Artist's sketch of the 18 lifts that carried the Vasa about 600 meters. Progress was not as even as this suggests, but it does show where the ship had to be turned around for easier handling. Note that she was facing away from Kastellholmen when found.

Air view of the last stage of the lifting operation off Kastellholmen. The vessels are, left to right, the Atlas, the Sleipner, the pontoons Oden and Frigg, and the Belos. Missing from the scene is the other salvage ship, the Ajax.

relieved. It was now believed that the *Vasa* could be brought to the surface intact since she had weathered so many dangerous moves.

To facilitate the whole operation and give it impetus, the Vasa Council was formed with His Royal Highness Prince Bertil as its Chairman. The Chairman of the City Council, Carl Albert Anderson, became its Vice Chairman and many prominent people were appointed as members. One of them was, of course, Anders Franzén. It was up to this Council to make further plans for the future of the *Vasa*.

Many more suggestions were now received by the Council as to how to bring the *Vasa* to the surface. One was that the ship should be filled with ping-pong balls, the theory being that the buoyancy of the balls would raise her. Another idea was to freeze her into a solid mass of ice, which would then float to the top. All suggestions were taken under advisement, but conventional methods prevailed, as before.

The Council also would be responsible for procuring funds. A modest estimate of the cost of raising the *Vasa* and building a museum was between one and a half and two million dollars, a sum which the Council hoped to raise from government and industrial sources and private donations. By May, 1960, the government and industry had contributed sufficient funds so that "Operation Vasa" could proceed.

During the summer of 1960 a great deal of work was done on the hull before the final risk of surfacing her could be taken. She had to be made as watertight as possible to float her and later prepare her for being pumped out after the surfacing. Most of this had to be accomplished while the ship was resting on the bottom. Gun ports were sealed off with temporary wooden lids, and all the holes left by the bolts as they rusted and fell off had to be plugged. Many heavy construction pieces were now removed to lighten her by 200 tons. The ragged edges of the poop deck were planked in to strengthen the whole stern.

Technical difficulties prevented the pontoons *Oden* and *Frigg* from lifting the *Vasa* by the same method as had been used underwater. Thus they were rebuilt during the winter of 1960 and re-

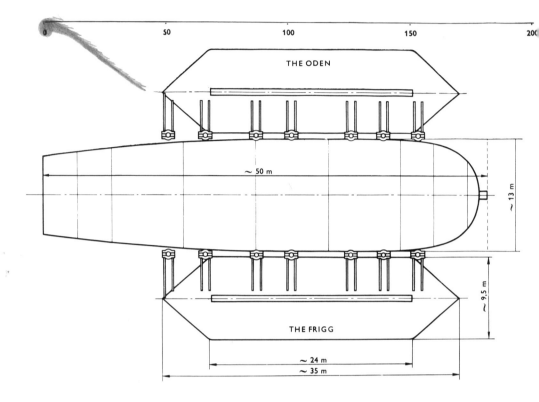

This diagram shows the Vasa *ready for the final lift. She is between the* Oden *and* Frigg, *now each equipped with seven hydraulic jacks.*

turned to their stations over the *Vasa* in the spring. Seven hydraulic jacks had been added to the facing sides of the pontoons, each with a lifting capacity of 50 tons — sufficient to lift the 500-ton ship. The original double cables were replaced by single nine-inch wires, strung under the keel, one end attached to a jack on one pontoon and the other end to the corresponding jack on the other pontoon, again forming a cradle. To lessen the strain of the heavy stern that extended beyond the pontoons, two pairs of submersible rubber pontoons were placed under the rear of the keel.

Finally, in April, 1961, all the problems connected with this stage of the operation were resolved and the crew was ready to make the final lift. The public and press had been alerted to the momentous event.

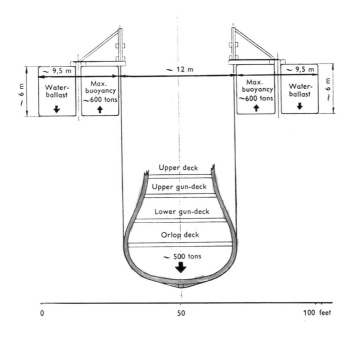

A cross-section of the final lift.

On the 24th of April, representatives from all over the world gathered to witness this feat — in boats, on the shore and in helicopters in the air. Over three centuries had passed, but the feeling of expectancy was reminiscent of that fatal day in August, 1628. The atmosphere was tense and hushed. Would the *Vasa* be able to take that long last leap? Or would she protest this tampering with her solitude and fall apart? At last the crew started the mechanism and slowly the jacks began hoisting their heavy burden to the surface. As the first stanchion appeared out of the calm waters, wild cheers broke out, flashbulbs popped and boat horns tooted. The *Vasa* had made her second debut! This glorious moment was recorded by television, radio and the press for the whole world to share.

The dramatic performance over, the crew hastily returned to their work, for the *Vasa* was not yet safe. With her top planks barely above water, pumps were brought aboard. Now began the tedious work of pumping the water out of the hull, preparatory to

The Vasa *breaks the surface on April 24, 1961.*

moving her into dry dock. While this was going on, divers and frogmen kept on plugging the innumerable holes to make the sides water-tight. Above board the beams and stanchions were wrapped in a plastic material as they were exposed, to prevent the wood from drying out. They would stay this way until the preservation phase began.

It took until May 4th, with pumps now working on all decks, to empty her of enough water so that she could be moved into Gustav V dry dock at Beckholmen, a short distance away. The happy and proud crew, headed by Captain Axel Hedberg, now eased her into her new berth, floating on her own keel, but listing to port because her stone ballast was still in its awkward position. Here her old friends, *Oden* and *Frigg*, and the other salvage vessels left her for good. The great ship *Vasa*, had finally come home.

On May 4, 1961, the Vasa *moves into the Gustavus V dry dock on Beckholmen, floating on her own keel. Note the slight list to port, caused by her uneven ballast.*

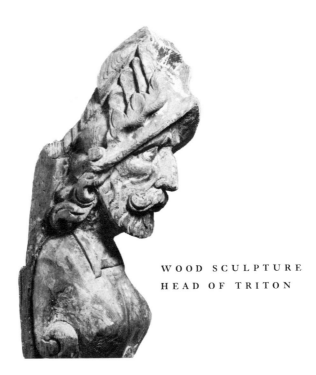

WOOD SCULPTURE
HEAD OF TRITON

5. *The Preservation*

WHEN the *Vasa* was safely inside the dry dock, it was
sealed off and the water pumped out except for eight feet in which
the ship rested. Now all the plugs and other sealing material put in
by the divers under water were removed to help speed the receding
waters. To counteract the list to port, several tons of scrap iron had
been attached to the ship's starboard side. Shortly after entering
dry dock, she was moved a ship's length by small submersible pontoons to the top of a concrete raft, measuring 184' x 60' x 12'. This
had been especially built for her and would be her permanent base.

All this time she was being sprayed continuously with water to keep her from drying out.

In the meantime, when the top deck was clear enough, the marine archeologists boarded her. Under the supervision of Per Lundström, they started the excavation of the interior. They subsequently set up shop on the upper gun deck and here the finds were brought from the mud-packed interior to be recorded. Most of the objects, except the ones that would have been destroyed by the lifting operation, had been left in the hull, since it had been impossible to dig them out beforehand. These were then hosed down and brought to a special preservation house, set up nearby for that purpose, where a most important phase of preservation for each piece was to begin, supervised by Lars Barkman.

After the discovery, the lifting and the moving, most of the interest in the ship had been of an archeological nature. Little had been done about the preservation aspect except for a few tests. Many people were of the opinion that no special treatment was required, since the ship was built of strong oak. Others felt that it would be a useless and expensive task — nothing could keep her from drying out and ultimately decaying. To resolve the matter, the Vasa Council, realizing that preservation would be a necessity, formed a Preservation Committee. It consisted of experts, each one able to analyze one or more of the materials found on board. This committee met regularly and probed into various methods of preservation. After much study and discussion, it decided upon the final preservation technique.

Most of the problems arising from the preservation of the *Vasa* were new ones, since there had been no previous marine archeological find of this magnitude. Thus these scientists had to start from rock-bottom. Their first and most important problem was the wood of the hull. True, the wood was well-preserved because (1) the *Vasa* was a brand-new ship when she sank; (2) by settling into the mud of a well-protected harbor, she had not been exposed to the wear and tear of the currents of the open sea or, as Anders Franzén had predicted, the ravages of the *teredo navalis*. Upon close examination, no evidence of this worm's activity could be found, and thus

the *Vasa* was structurally sound. This, then, left the problems of shrinkage and rot.

Much of the *Vasa* was made of heartwood, the innermost core of the oak. This normally would be unaffected by rot because of its hardness. However, there was some rot, which had to be stopped. It was also found that the wood had been affected differently according to its position. The upper parts of the hull, the trimmings and the undersides of the decks showed signs of "soft rot," while the entire lower part of the hull and the tops of the decks were

OPPOSITE TOP: *Archeologists cleaning the lower gun deck with water. At left is a crude but sturdy gun carriage, still in position.*

OPPOSITE BOTTOM: *An archeologist examining one of many chests found on the orlop deck. A broken barrel is in foreground.*

BELOW: *The lower gun deck, starboard side, after complete excavation. Light and air penetrated to the lower decks through the square-shaped holes in the center of the ship. The anchor cable was lashed around the horizontal beam in the left background.*

Lars Barkman, head of the preservation department, beside one of the tanks in the laboratory. He is checking a separator that cleans the preservative so it can be used over and over again.

strong and hard. The conclusion drawn from this observation and from microscopic tests was that the parts submerged in the clay and the top of the decks covered by mud had been preserved by this matter, while the exposed areas had been attacked by mold-producing fungi (*ascomycetes* and *imperfecti*). The damage from the fungi however, was slight — only one or two centimeters of the outer surface. Attacks on the inner parts were attributed to bacteria and/or chemical decay.

Another interesting observation was that the dark, discolored areas around the holes and joints in which iron bolts had lodged were very strong. This proved that the rust, left by the bolts that had fallen out, had helped preserve the surrounding wood. Sections made of softer woods, like pine and birch, or those not rusted, did not fare as well, for they were almost totally destroyed.

The Committee realized that the fungi problem would be the easiest to solve. After much research, it was decided that the water content of the wood fibers could be kept at a level that would discourage fungi attacks. As an extra precaution, a mixture of boric acid and borax in a 7:3 ratio was added to the water in the preservation tanks. This guarded against both insects and fungi. Tests taken over the years have proved this method effective.

Shrinkage was a much more serious matter. A water-soluble, somewhat stiff material had to be found to replace the water now saturating the wood cells. This material, when enough had penetrated the fibers, would reinforce them and prevent cracking. Before proceeding with research on such a material, tests were conducted to determine how much the hull had swelled and thus, in turn, establish the amount of shrinkage necessary to restore it to normal.

The Vasa on her permanent concrete pontoon in dry dock, being sprayed with salt water. The beams and stanchions have been covered with plastic to keep them from drying out.

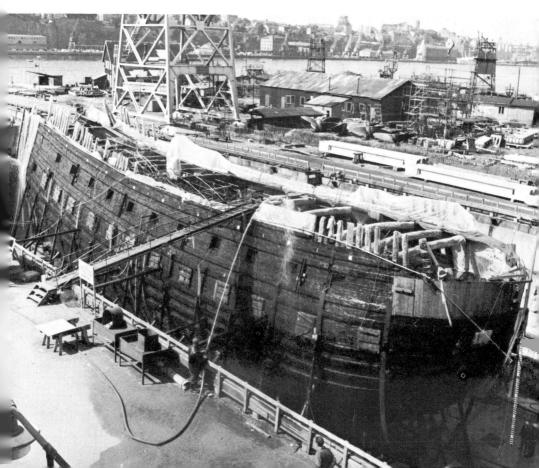

A piece of Vasa *oak that had been allowed to dry out naturally.*

The results showed that there was very little swelling lengthwise and only about 1% in the width overall. In another test, a piece of oak taken from the *Vasa* was allowed to dry naturally, without exposure to direct sunlight and without artificial drying. The results showed at least 15% shrinkage, proving conclusively that another means of drying the hull must be found soon. Otherwise the whole ship would have to be dismantled and preserved in pieces to be restored at a later date when time and the economic situation permitted.

Until the right preservative could be found, the *Vasa* was kept in dry dock in the open. This was the rest of the spring, summer and into November. Meanwhile she was sprayed continuously on the outside surface with sea water — at an average of 4400 gallons per minute — a process that also cleaned off a good deal of the mire that covered her. Because of the archeologists working in the interior, only fresh water was used to clean out these areas, much to the chagrin of the local inhabitants whose water pressure was lowered at the same time.

In the meantime the loose sculptures at the original site were continuously being brought to the surface. This operation finally was finished at the end of summer, 1967. These finds were all in various degrees of decay, depending on the wood and resting place. The oak pieces were in fairly good condition as were the pine. The lime, birch and other softer woods had deteriorated more, as had the objects still in place on the upper deck. They had been exposed not only to attacks of fungi but also to the abrasions of the currents and the anchors of other ships. The sculptures were put in special vats in the preservation building containing pure water and a solution of the rot-preventative chemicals. As far as possible, pieces of the same wood and size were put into the same vats, since their preservation would take the same length of time.

After much experimenting, the scientists finally found the "stiff material" for which they had been searching. It proved to be a mixture called polyethylene glycol,* which hereafter will be referred to as PEG. It had a melting point of 87° Fahrenheit. The temperature of the water was raised to 92° Fahrenheit to insure that the solution would mix with water and penetrate the wood fibers. This period of preservation took from between one to two years, depending on the wood and the size of the object. During this time the concentration of PEG was gradually increased and the temperature raised very slowly. This was to keep the wood from absorbing the solution too rapidly, thereby expelling the same amount of water too soon with the danger of shrinking in the vat. This process could not be rushed after all those years of being waterlogged.

After the immersion period in the vats, the drying period started. During this the surface was treated continuously with the glycol mixture. This took from six months to two or three years, depending upon the size of the piece. When the complete cycle was finished, the sculpture was either exhibited in the museum or stored, awaiting the time when the full restoration will take place.

The other finds were treated in a similar manner. A hat and some articles of woolen clothing were cleaned and treated with a strengthening solution, containing a special type of glue. Leather finds were treated with the PEG solution.

---

* Invented by Union Carbide, U.S.A., the American brand name is "Carbowax." It was patented in Sweden for the preservation purpose by Rolf Moren and Bertil Centerwall to MO and Domsjö AB.

Six sails that had never been hoisted were found in an orlop-*
deck locker together with the mizzen bonnet and sail for the ship's
boat. They were badly deteriorated and needed a more complicated
procedure. After a careful cleaning, ethyl-alcohol and xylene were
applied to the fabric and then dried out slowly. If the sails had
dried out naturally without this treatment, they would have fallen
apart when lifted. After they were dry, they were placed on sep-
arate pieces of framed glass-fabric, cut to the shape of the original
sail. This glass-fabric had already been treated with a polymer solu-

* See cross section of the ship on page 74.

A felt hat after treatment. Found in a sailor's chest, it was of a style common during the 17th century.

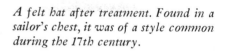

Chain shot — one of the restored iron objects.

FAR LEFT: *Passa glass. This green-hued octagonal wine glass was imbedded with wire rings that measured the amount each man could drink as the glass was passed around — hence its name.*

LEFT: *A modern replica of the passa glass is sold at the Museum. The rings, however, are glass.*

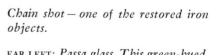

The preserved sail from the Vasa *longboat, on display in the lecture hall.*

tion. The same solution was then applied to the sails so they would adhere to the backing. When dry, this solution became an invisible plastic coating so that the original sail could be clearly seen.

Other methods were employed to preserve the glass, pottery, pewter and iron finds. The iron objects had rusted to a very advanced degree, the wrought iron more so than the cast, which had maintained its original volume and shape, as, for instance, the cannon balls. If these and the other iron objects had been allowed to remain in the air in the condition found, they would have completely disintegrated from oxidation. A special furnace was installed in January, 1965, to preserve the iron objects. Placed inside, and with hydrogen gas added, they were heated to 2247° Fahrenheit. This process reduced them back from iron oxides to pure iron.

As the work on the wooden sculptures progressed, it became clear that PEG could be used in preserving the hull.* Before this process could be started, the entire exterior and interior had to be

* A lower molecular weight was used — 1500 PEG as compared to 4000 PEG for the sculptures.

[67]

thoroughly cleaned. The latter proved to be a most difficult job, because it involved getting into all the odd angles and inclosed spaces between the hull and the inner walls. In some cases, planks had to be removed and special tools constructed to reach the filth.

During the cleaning, a ferro concrete framework was built around the *Vasa* on top of the concrete pontoon on which she was resting. Then she was towed out of dry dock into undisturbed waters nearby, where her boat house, consisting of aluminum sidings and roof, was added. This was to be her temporary housing while she was being preserved. Finally, in November, 1961, she was towed again and secured to the shore at the site where the Vasa Museum had been established.

Now began the preservation of the hull, undoubtedly the largest and most ambitious project of its kind ever attempted. It was sprayed by hand once a day in this way: the preservative, enclosed in a tank holding 660 gallons, was pumped through a pipeline attached on three levels both on the inside and outside of the ship,

The Vasa *being towed to the spot where her boathouse was added.*

Two views of the hull being sprayed.

OPPOSITE: *The* Vasa *inside her temporary boathouse during one of the 20-minute spraying periods.*

and fitted with hose connections at regular intervals. Five men systematically went over the whole ship, a procedure that took five hours. After each treatment the entire piping system was cleaned out with compressed air. Between sprayings, a humidifying system maintained the level of 95% humidity at all times to keep the wood from drying out and to secure maximum penetration of the preservative.

The effectiveness of this method was measured from April, 1962 to February, 1965. The results showed that the PEG had penetrated to a satisfactory depth of 7 inches, but the wood had not retained enough of the PEG to preserve it successfully. Therefore a fully automatic spray system was installed, and since March, 1965, the entire surface has been sprayed continuously during the hours that the museum is closed. All through visiting hours it is sprayed every other 20 minutes. The public can witness this damp and noisy event or wait until the next lull. This system has proved more economical than the first method, since only one person is needed to operate it and the liquid is recirculated instead of being flushed away.

It has also proved more satisfactory from the standpoint of PEG retention. It is expected that the spraying will be continued at this level for at least another year. Then the plan calls for spraying the surface twice a day for another two years. After this the wood will gradually be dried out during which time the moisture ratio will be stabilized.

The entire process is expected to take many more years before the ship can be pronounced completely preserved. In the meantime, certain construction details are gradually being added so that little by little the *Vasa* is assuming her old shape. One of these is the lion figurehead, which once again is perched on the prow, proclaiming the might of the *Vasa*.

6. The Ship

WHEN the *Vasa* was raised and moved into dry dock, all questions about her appearance could gradually be answered. Up to now there had been only conjectures, since, previously mentioned, no drawings of the ship existed. Shipwrights did not make plans before the 18th century. Before and while constructing a ship, workmen were sent out into the forests to search for trees that had the natural shapes of frames, angle timbers, braces, deck beams and other parts. The pieces were then cut from these trees according to specifications given to the workman orally by the shipwright who had them stored away in his mind, not on paper. However, there was some tangible information about the *Vasa* left for posterity. Records, dating from 1625 and 1626, showed some purchases of wood cut by workmen; also other purchases, mostly oak planking, waling and cambered oak, and a variety of other woods.* The local oak was owned almost exclusively by the Swedish nobility and from their forests came the oak with which the *Vasa*'s hull was built.

As items made from the various woods were brought up to the surface, little by little a true picture of the ship's superstructure began to emerge. It was noted that the upper gun deck had smaller gun ports but that both decks had the same size guns. Of these, 48 were 24-pounders, 8 — 3-pounders, 2 — 1-pounders and 6 mortars. Only three guns were actually found aboard the *Vasa*. The others, brought up by the early salvagers as previously described, have disappeared, their whereabouts now shrouded in mystery. Some cannon balls were recovered, as were bar shot, chain shot, spike shot and six kegs of lead musket shot. Since only a few wooden parts of hand weapons were found, probably belonging to the officers, the theory that the soldiers had not boarded the vessel seems to be substantiated. They would have brought at least one gun apiece, which would add up to a large number with the following complement of 437 men:

3 Officers
12 Non-commissioned Officers
12 Artisans or Specialists
90 Ordinary Seamen
20 Gun Captains
300 Soldiers

* These were alder, ash, beech, linden, spruce, pine, maple, willow, walnut, apple and pear, some of which came from Holland and the Baltic provinces around Königsberg and Riga.

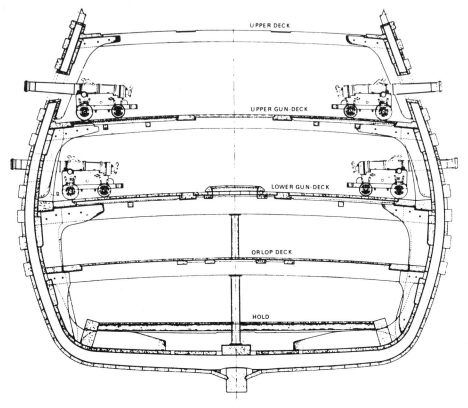

UPPER DECK

UPPER GUN-DECK

LOWER GUN-DECK

ORLOP DECK

HOLD

Cross-section of the Vasa, *showing the upper deck, upper gun deck, lower gun deck, orlop deck and hold.*

Longitudinal drawing of the Vasa. *Note the slant of the mainmast. Immediately to its right, between the orlop deck and the hold, is the galley (u-shaped) made of brick.*

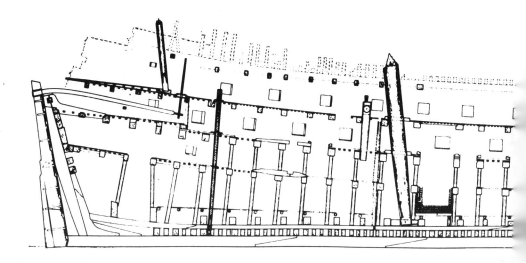

Now it was ascertained that the *Vasa* measured 170 feet from keel to masthead. Her total length, including bowsprit, was about 230 feet — 165 feet from stem to sternpost. Her beam was 42 feet and her original displacement just over 1300 tons. An interesting observation — her ballast of rocks was 120 tons, astonishingly little for a ship her size, according to modern standards.

The *Vasa* had three masts, the mainmast estimated as being between 180 and 190 feet high with a diameter ranging from 30 to 40 inches. It had been built with a center square section and four outer curved pieces to make it round. The rudder, which was found fully intact attached to the ship, was brought to the surface in 1959. It was 30 feet in height and very narrow compared to modern ones. This was because the rudder in this type of ship was used only for minor adjustments in steering. The important factor in setting the course was the handling of sails, of which the *Vasa* had ten: foresail, fore-topsail, main-topsail and mizzen, all four lost when the ship went down, and royal spritsail, mainsail, fore- and main-top-

Erik Åkerberg's sketch of a similar ship, showing how the rudder was controlled from the main deck.

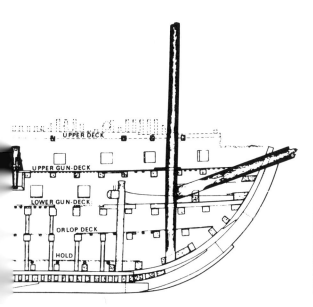

gallants and mizzen topsail, found in the orlop locker with the mizzen bonnet and the sail for the ship's boat. Evidently, extra sails had not been brought along.

Speculation that the *Vasa* had been richly decorated was confirmed as the wooden sculptures were brought up by the divers. By the 17th century, wood carving had become a fine art. What better way to beautify a warship of the royal fleet than to adorn her with a generous sampling of the wood carver's skill? Dominating and leading all other ornaments in magnitude and magnificence was the lion figurehead, already mentioned, and brought up in 1959. Carved out of lime wood, it weighed two tons and was ten feet nine inches long. Because of its tremendous size, it had been made in three sections, as had many of the larger figures. It still showed traces of gold leaf and paint after it was washed.

The preserved lion figurehead attached to the bow, which is now undergoing reconstruction.

The lion was often used as a figurehead. As a symbol of strength and power, it was a fitting herald for a ship of the House of Vasa. A lion face, too, adorned each gun port lid. Flecks of red paint clung to them. From chemical analysis of these and other objects, it has been presumed that all ornaments were either gilded or painted in red, yellow and brown.

Besides the lions, there were carvings of soldiers, mythological creatures, musicians, mermaids, caryatids, cherubs, grotesque heads, heraldic devices and other miscellaneous subjects. These figures were in a style similar to the baroque, rich in detail and decorative quality, typical of continental Europe in the early 17th century. This art, besides being decorative, is full of symbolism. Each statue or object connotes royal power and wealth, mythological deeds, or man's sinfulness and eventual death.

FAR RIGHT: *Wood sculpture of a musician from the aftercastle.*

RIGHT: *Wood sculpture of the Greek god Hercules, enfolded in a lion's skin, with the dog Cerberus chained at his feet.*

Some decorations were supposed to have a psychological effect, such as the lion faces on the gun ports. They were to scare off the enemy, while, on the other hand, the lids, as well as the decks, were painted red to make the sight of blood less terrifying to the crew.

Similar examples of both the woodcarver's and the stonecutter's skill can be found in many buildings of old Stockholm, dating from the Renaissance period. A number of these artists worked on the *Vasa*, for such a gargantuan project required all the artisans available. Their names have been found in the *Vasa* records. Three have proven of special interest: a German, Martin Redtmer; a Dutchman, Johann Thesson; and Hans Clausink from Westphalia. Professor Sten Karling, noted art historian and expert on the *Vasa* sculptures, tracked down a number of their architectural carvings and other sculptures. By comparing certain characteristics with those of the ship's pieces, he was able to determine which were undoubtedly the work of these three men. Notable was the statue of Hercules, whose face closely resembled the face of a statue, known to be Redtmer's work. Erected in 1647, it is still standing in the main square of Stockholm's Old Town. A copy can be seen in the Vasa Museum.

The *Vasa* sculptures were mostly clustered in two distinct areas, the aftercastle and the forecastle, but others were scattered throughout the ship. A number of small Hercules figures adorned the poop deck gunwales, and there were carved balustrades fore and aft. Carved panelling and molding decorated the captain's cabin. The main piece of the forecastle's group has already been mentioned — the lion, lunging forward holding a shield between his paws. He was flanked on each side by a row of Roman soldiers and above them was a curved rail holding helmeted heads. The whole group was guarded on the port side by a seven-foot warrior in Roman armor, standing on a lion's head with a dog at his feet.

By far the greatest amount of sculpture belonged to the sterncastle. This had double quarter galleries, the lower and largest leading to the captain's cabin. A description of this entire section can be found in the caption on the opposite page. The symbolism of

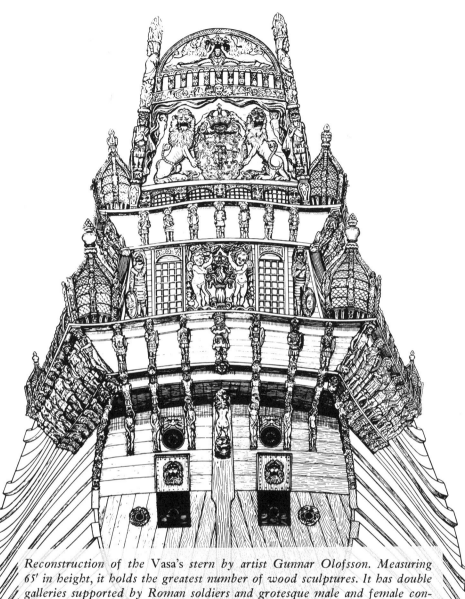

Reconstruction of the Vasa's stern by artist Gunnar Olofsson. Measuring 65' in height, it holds the greatest number of wood sculptures. It has double galleries supported by Roman soldiers and grotesque male and female console heads. The roofs of these galleries and the corner cupolas are decorated with reclining mermaids and tritons. The upper part of the sterncastle consists of a huge wall bearing two lions holding the royal coat of arms. Separating the two windows in the lower gallery is a panel with two cupids holding the Vasa sheaf, the symbol for the ship. Barely visible on the top cross beam are the initials GARS (Gustavus Adolphus Rex Sueciae) on either side of a winged head.

all these figures is noteworthy — for instance, the three crowns in the coat-of-arms represent the three wise men; the crowned wheat sheaf, the House of Vasa and, thus, the ship's name; the bizarre console faces, man's helplessness and his tortuous path to corruption and death; the Roman soldiers, the strength of the Roman emperors, from whom European monarchs believed they had descended. It is interesting to note that the King saw fit to draw his strength and inspiration from the pagan Romans rather than the Christian God for whom he was waging his thirty-year war.

With these structural finds and works of art, the *Vasa* will someday be reconstructed and, it is hoped, will look as fine as the day she set sail.

But what about the human element? What has been discovered about those who manned her? Of the 24,000 listed finds,* most were, of course, construction details, but many were personal items belonging to the crew. From these the archeologists were able to learn a great deal not only about life aboard a warship but also about the peasant in the 1620's. This is very important because, although museums all over the world display the wealth that belonged to the aristocracy of the era, very little survives that gives us an insight into the poor man's lot. Thus these articles have proved to be an exceptionally valuable archeological find.

Each man had to provide his own eating utensils as well as food for his first meals, since no ship's food was served until after the ship had sailed through the achipelago. The eating paraphernalia consisted of a small earthenware bowl with a looped handle, a round wooden dish and a sheath knife. If this was not enough equipment, a man used his fingers. Many personal items were found in barrels, chests and ditty-boxes, as were large earthenware bowls used for cooking, serving and storing food. Most of these were red-fired, lead-glazed ceramic ware and some were decorated. Once at sea, the crew ate directly from these bowls. A number of men

---

* Of the registered finds, there were 14,000 construction details and almost 800 sculptures and carved details. The divers retrieved 3500 pieces.

ABOVE: *A bottle and bowl with looped handle were two of the earthenware objects found aboard the* Vasa.

ABOVE RIGHT: *A sailor's small wooden chest, one part of a wooden box, a three-legged clay pot and a piece of rope.*

RIGHT. *A sailor's ditty-box that contained, among other things, a wooden spoon, a ball of twine and the copper öre at left.*

Set of carpenter's tools found aboard the Vasa.

were assigned to each one and shared its contents. The fare was simple, consisting mostly of dry, cold food, such as various grains and salted fish and meat. Because of the fear of fire, hot meals were prepared in large iron pots only when the sea was comparatively calm. These cooking pots were suspended over an open fire on the brick hearth in the center of the orlop deck.

The tableware thought to belong to the officers was a little more affluent-looking. There were pewter tankards, flasks and plates, most carrying still legible Swedish hallmarks so that the craftsmen have been traced. A blue Dutch delftware plate, a bronze candlestick, a bronze brazier and a passa glass were among the more luxurious objects found.

Blue delftware plate from Holland.

The pewter brazier, plate, tankard and bronze cock belonged to the officers.

One of the pewter tankards contained a liquid, which, when analyzed, was found to be 33% alcohol and similar to contemporary rum. A wooden box held butter, just a little rancid. Vegetable fragments proved to be corn, oats, peas, rye, wheat and seed shell, probably beets or spinach. Other fragments analyzed proved to be animal fat — probably the minute remains of the beef, pork and fish, which had been salted away in barrels, A piece of veal shank was also found.

Other objects of interest were chests with carpenter tools, cruder than today's but still similar to modern ones; an apothecary kit, consisting of mortar and pestle, one short and two long-handled spoons, a pewter flask, a stoneware jar, a wooden frame, a whisk and a metal cap shaped like a cock. This kit must have belonged to one of the crew who acted as the ship's physician. If it was his complete equipment, it was meager indeed. What he used for medicine is not known.

No precious metals were uncovered except for 74 silver coins, a gold signet ring with its stone missing and a silver spoon that had traces of ornamentation. However, thousands of copper coins were found in different locations aboard, suggesting that each of the

The captain's table laid with some of the more precious objects. The flask at right contained a liquid similar to rum.

crew had at least a small supply. (This is the largest single collection of copper-minted coinage ever discovered in Sweden.) These coins were mostly one-öre pieces from the Falu copper mines in Dalarna in central Sweden. It was this mine that, to a great extent, financed the war King Gustavus II Adolphus was waging at this time. Some of these öre, called *klippingar*, were square-shaped in order to cut down on production costs. An öre today is worth about .002 cents. In the 1620's a hen could be bought for 24 öre — less than five cents. Woe to the culprit who dared to steal an öre. His punishment was death!

Many articles of clothing in different degrees of decay were also found — some in trunks, some on or beside the skeletons recovered. Examination showed that there was no particular uniform for the crew. Each man wore his own everyday clothes. One wide-brimmed hat, made of felt, was in very good condition, as were a leather boot and same low leather träsko, a slipper-type shoe, still popular today even in America.

Probably the most poignant finds were the skeletons or parts of skeletons belonging to eighteen individuals. Anthropological studies of these determined that at least two were of women between 25 and 30 years old and the rest were of men between 25 and 50. One woman was about 5′ 2″, the other about 5′ 5″, while the men ranged from just under 5′ 4½″ to 5′ 7″. We might surmise that the average man was not very tall in this era, so he would fit, not too comfortably, into the low quarters below deck, his permanent station for the journey.

From cranial studies, two different ethnic types emerged. The remains of one man and one woman showed they had short skulls and prominent cheekbones. They were thought to have come from Finland. The others were identified as Nordic types, native to Sweden. In one skull were the remains of a brain, recognizable as such, although greatly shrunk. One male had his right shinbone broken in two places, presumably by the impact of the accident. Another showed a healed cut wound on his left shinbone and thigh, and still another showed symptoms of syphilis.

In 1963, after all necessary studies had been made, these individuals were buried at the Naval Dockyard Cemetery in Stockholm at a special service. Thus finally ended, for those poor souls, the journey which began so illustriously in 1628 and ended so ignominiously 335 years later.

Now that most of the pertinent facts about the *Vasa* have been brought together and a clear picture of her has been formed, we, too, can question the reasons for the *Vasa*'s sinking, just as the people of Stockholm did so long ago. We can ask and try to answer the same questions:

1. Was the *Vasa* built too narrow and sharp without enough belly and ballast?
2. Was she top-heavy?
3. Were the lower gun ports too close to the water?
4. Were the guns at fault — too heavy and misplaced?
5. Will there ever be a definite answer to this mystery?

There are still many different theories about why the *Vasa* sank. But the experts agree on the following facts. The *Vasa* was built too narrow. Her ballast was only 120 tons of stone although a ship her size should have had 400 tons. But if the *Vasa* had had that ballast, the water line would have been above her lower gun ports. The *Vasa* was extremely sturdy and well-built to have survived her long emersion, but that enormous sterncastle did help to make her top-heavy. Furthermore, the same heavy guns — 24-pounders — were used on both the upper and lower gun decks, although the top gun ports were smaller indicating that lighter guns should have been used.

So, according to modern science, the answers to our first four questions would be "yes." As to the fifth, perhaps a definite explanation will emerge when the *Vasa* has been completely restored and reconstructed.

7. *The Museum*

T HE present buildings of the Vasa Museum* are only temporary, and are therefore called the "Wasa dockyard." A permanent

* The museum still bears a sign showing the "Wasa" spelled with a "W," which
was used in the beginning because of the decision made by the temporary Wasa
Council. Some time later, the Royal Swedish Academy and the Advisory Council
for Swedish Terminology and Usage recommended that the *Vasa* be spelled with
a *V* rather than a *W* to conform to the modern Swedish alphabet (Latin) that has
no *W*. Hence the use of *V* throughout this book.

structure will eventually be erected to house the completely re-stored and reconstructed ship.

In the meantime a visitor to the museum can get a very good picture of the *Vasa* and what life aboard her was like from the exhibits. Directly beyond the entrance gate is Anders Franzén's reconsrtuction of von Treileben's diving bell. Behind it two bronze cannon are pulled up in back of lion-adorned, open gun ports, cut in a reconstructed section of hull. In the center of the courtyard beyond, rests one of the *Vasa*'s anchors. Then comes the aluminum boat house that encloses the ship's hull, still on its concrete pontoon in the water.

To the right is a long wooden two-story wing with executive offices on the second floor and an exhibit area on the first, guarded by Edvin Fälting's diving suit, standing at attention. Here various objects from the ship are displayed — such as bowls, tankards, money, tools, shoes, a hat — with pertinent information alongside. There is a model of a sailor wearing a composite of various articles of clothing found on the different skeletons. Here, too, is Anders Franzén's core sampler and models of three different phases of the salvage operations, donated by the Neptun Salvage Company.

Front view of the Vasa Museum. The Vasa *is housed in the building at the rear of the courtyard.*

LEFT: *This figure wears a composite of the clothes found on the skeletons. The sailors had no uniforms but wore the street clothing of the period.*

OPPOSITE TOP: *Top view of one of three models that show different stages of the salvage operation. A compass at top right, shows that the* Vasa *was headed almost due east.*

OPPOSITE BOTTOM: *Side view of same model, showing the* Vasa *in her cradle before the first lift. Compare the size of the* Vasa *with the diver standing by the stern.*

Another wooden wing on the left houses poster exhibits of the foods brought on board. The wooden sculptures that have been through the preservation process can be found here along with some of the pieces described in chapter six, such as the Roman soldiers, cherubs, caryatids, grinning console heads, and many more. Here, also, is the replica of the Martin Redtmer statue, located in the "Old Town," that gave a clue to the identity of one

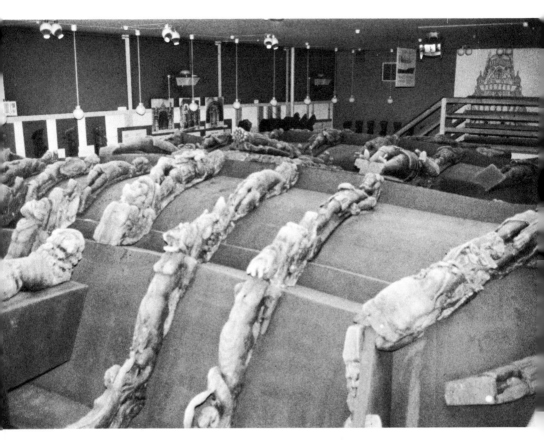

Reconstruction of the sterncastle, showing reclining figures.

of the *Vasa*'s sculptors. A most interesting exhibit is the reconstruction of the stern with two gun ports, the rudder and most of its sculptures in position. Because of its great height, the stern has been constructed flat on the floor instead of in a vertical position, with steps and a platform over it so it can be viewed from above. The much debated water-line has been painted in its exact place.

In this wing there is also an auditorium where colored films, telling the story of the *Vasa*, are shown regularly in several languages, depending on the audience. More than three million people from all parts of the world have seen the *Vasa* to date and each year about forty thousand children visit the museum where special programs are presented to them.

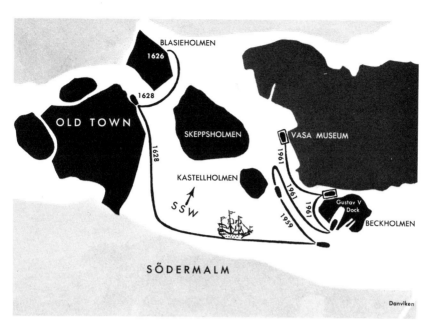

The complete voyage of the Vasa.

Another view of the reconstructed stern, showing gun port and water line.

Harbor view of the Vasa Museum. In the far distance, directly to the left, is Blasieholmen, where the ship's keel was laid in 1626.

Outside this hall is a souvenir shop where one can purchase exact replicas of many of the restored objects, plus books, prints, coins and commemorative stamps. A restaurant completes the wing.

The preservation is still going on in the original building a short distance away from the museum, on the island of Beckholmen where the ship was in dry dock.

The plans for the future of the museum are not definite and there is no immediate need to rush them, for the preservation of the hull will take some years more. No one knows what shape the final structure will take, but it will certainly be equal to housing this beautiful Queen of the Baltic.

Bibliography

BOOKS

Almquist, Bertil. *The Vasa Saga.* Stockholm: Bonniers, 1966.

Andersson, Ingvar. *A History of Sweden.* London: Weidenfeld and Nicolson, 1968.

Franzén, Anders. *The Warship Vasa.* Stockholm: Norstedts/Bonniers, 1960.

———. *The Warship Vasa.* 5th ed. Stockholm: Norstedts/Bonniers, 1966.

Negri, Francesco. *Viaggio Settentrionale.* Padova, 1700.

Ohrelius, Bengt. *Vasa, the King's Ship,* trans. Maurice Michael. New York: Chilton, 1963.

Triewald, Mårten. *Konsten att lefwa under vatn.* Stockholm, 1734.

Widding, Lars. *The Vasa Venture.* Gebers.

BOOKLETS, CATALOGUES, PERIODICALS

Barkman, Lars. *On Resurrecting a Wreck.* Swedish National Maritime Museum, Stockholm, 1967.

———. *Replica of Relics.* Swedish National Maritime Museum, Stockholm, 1969.

———. *The Preservation of the Vasa.* Swedish National Maritime Museum, Stockholm, 1965.

Franzén, Anders. "Ghost from the Depths: The Warship Vasa," *National Geographic* Magazine (January 1962), 42-57.

———. *Vasa, The Strange Story of a Swedish Warship from 1628.* Stockholm: Bonniers/Norstedts, 1962.

Kleingardt, Bergitta, ed. *Wasa.* Swedish National Maritime Museum, 1968.

Statens Sjöhistoriska Museum. Stockholm, 1960.

Index